# Education 141

# 化装舞会

## Dress Up Party

**Gunter Pauli**

冈特·鲍利 著

凯瑟琳娜·巴赫 绘
何家振 译

学林出版社
www.xuelinpress.com

## 丛书编委会

主　任：贾　峰
副主任：何家振　闫世东　郑立明
委　员：牛玲娟　李原原　李曙东　李鹏辉　吴建民
　　　　彭　勇　冯　缨　靳增江

**特别感谢以下热心人士对译稿润色工作的支持：**

王必斗　王明远　王云斋　徐小怗　梅益凤　田荣义
乔　旭　张跃跃　王　征　厉　云　戴　虹　王　逊
李　璐　张兆旭　叶大伟　于　辉　李　雪　刘彦鑫
刘晋邑　乌　佳　潘　旭　白永喆　朱　廷　刘庭秀
朱　溪　魏辅文　唐亚飞　张海鹏　刘　在　张敬尧
邱俊松　程　超　孙鑫晶　朱　青　赵　锋　胡　玮
丁　蓓　张朝鑫　史　苗　陈来秀　冯　朴　何　明
郭昌奉　王　强　杨永玉　余　刚　姚志彬　兰　兵
廖　莹　张先斌

# 目录

| | |
|---|---|
| 化装舞会 | 4 |
| 你知道吗？ | 22 |
| 想一想 | 26 |
| 自己动手！ | 27 |
| 学科知识 | 28 |
| 情感智慧 | 29 |
| 艺术 | 29 |
| 思维拓展 | 30 |
| 动手能力 | 30 |
| 故事灵感来自 | 31 |

# Contents

| | |
|---|---|
| Dress Up Party | 4 |
| Did you know? | 22 |
| Think about it | 26 |
| Do it yourself! | 27 |
| Academic Knowledge | 28 |
| Emotional Intelligence | 29 |
| The Arts | 29 |
| Systems: Making the Connections | 30 |
| Capacity to Implement | 30 |
| Fable is inspired by | 31 |

青蛙遇到了水蚤，他想起来曾经在池塘里和池塘周围看见过她和很多其他水蚤。

"很高兴又见到你。"青蛙向漂亮的水蚤打招呼。

"你是谁？我从来没有见过你。你最好离我远一点。"

A frog meets a nymph and remembers seeing her and many others in and around the pond.

"Good to see you again," the frog greets the pretty nymph.

"Who are you? I have never met you before. So I would appreciate if you keep your distance."

青蛙遇到了水蚤。

A frog meets a nymph.

你不记得我吗?

You don't remember me?

"你不记得我吗？我一直在这里游泳，在这个池塘里好几个星期了。来，再看看我。"

"我的记忆力很好，"水蚤回答道，"但我从来没在这里见过你这样的。"

"你是睡眠不足，还是缺少维生素呢？"

"You don't remember me? I have been swimming here, in this pond, for weeks. Come on, take another look."

"I have a very good memory," the nymph responds, "but you do not look like anything I have ever seen here before."

"Do you not get enough sleep, or are you perhaps lacking an important vitamin?"

"你看看,先生。"她回答,"你一直在背台词,企图与我搭讪。我告诉你,我从来没见过你。所以,请问你现在可以离我远一点了吗?"

"但是,是我呀,你的朋友,那只青蛙呀。"

"Look here, mister," she replies. "You have tried every line in the book to try and strike up a conversation with me. I tell you, I have never-ever seen you before. So would you please leave me alone?"

"But it is me! Your friend, the frog."

但是，是我呀，你的朋友，那只青蛙呀。

But it is me! Your friend, the frog.

你让我想起了恐龙……

you make me think of the dinosaurs ...

"最近我们这里唯一的朋友是蝌蚪。我们看着他们吃植物和藻类,而你长着典型的食肉动物的脸。你甚至让我想起了恐龙。"

"这么说你的确认识那些蝌蚪,对吧?我就曾经是他们中的一员!"

"The only friends we have had around here lately were the tadpoles. We watched them eat plants and algae, but you have the typical face of a carnivore. You even make me think of the dinosaurs."

"So you do remember the tadpoles? I was one of them!"

"那又怎样?你假装是一只蝌蚪,一只只有前腿,有胡须、鳃片和一个螺旋形的嘴的小蝌蚪,然后一夜之间变成了一个有强壮的后腿、巨大的颌骨和一条长长的舌头的动物,准备好了吃我。"

"但是,我就是那只蝌蚪呀!"

"算了吧,我以前也去过化装舞会,很多人外表变化惊人,但我可从没见过一个外表变化像你这样大的。"

"How about that? You pretend that a tadpole, one with only front legs, who has whiskers and fins, and a spiral-shaped mouth is transformed overnight into an animal with strong back legs, a big jaw and a long tongue – ready to eat me."

"But it is me!"

"Well, I have been to dress up parties before, ones with many surprises, but never-ever have I seen anyone change his appearance like this!"

你假装是一只蝌蚪!

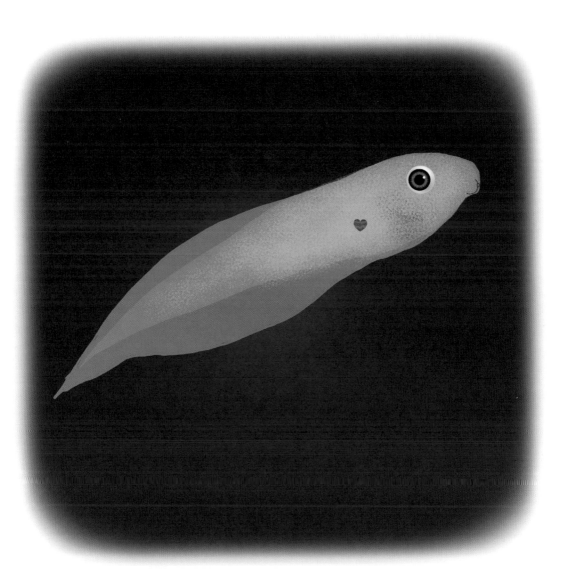

you pretend you where a tadpole!

所有的青蛙一开始都是蝌蚪!

All frogs are tadpoles first!

"但是，这是真的。所有的青蛙一开始都是蝌蚪，然后它们就变成了　　　"

"食肉动物。当你是蝌蚪时，你主要吃的是植物，但当你是青蛙时，你就会想吃我。"

"我的小水虿朋友，你的外表也会变化，可能比我的变化还大。几个星期后，你会变成什么呢？"

"But it is true. All frogs are tadpoles first and then they become ..."

"Carnivores. When you are a tadpole you eat mostly plants, when you are a frog you will want to eat me."

"Well, my little nymph friend, you will change your looks as well. Perhaps even more than I do. What will you be in a couple of weeks?"

"我会变成一只蜻蜓。"她骄傲地回答。

"你？你会变成一种地球上少有的史前昆虫，甚至在恐龙在地球上漫步之前就到处飞翔？"

"就是啊！但是作为水虿，我住在水里。"

"我知道。我以前常和你一起玩呢。"

"I will be a dragonfly," she says proudly.
"You? You will turn into one of those prehistoric insects that lived on earth, flying around everywhere, from even before the dinosaurs roamed the Earth?"
"Exactly! But as a nymph I live in water."
"I know. I used to play with you."

我会变成一只蜻蜓

I will be a dragonfly

我用屁股呼吸！

I breathe through my butt!

"当我准备离开水面时,我就会蜕皮,变成一只蜻蜓。"

"我一直很好奇,你是如何能够在水中游动的。你们没有翅膀,没有鳃,也没有腿。"

"那是因为我用屁股呼吸。"

"你开玩笑吧?没人用屁股呼吸,更不会利用屁股呼吸推动自己前进。"

"When I am ready to get out of the water I will shed my skin and turn into the dragonfly."

"I always wondered how you are able to move around in the water. You don't have wings, or fins or legs."

"That is because I breathe through my butt."

"You are joking! No one breathes through their backside, let alone making use of that to propel themselves forward."

"嘿，小青蛙，你要学习的东西还多着呢。我告诉你，我把水从我的屁股吸进，分离出我需要呼吸的空气，然后排出水，这样就能推动我前进。"

"好吧，如果你想在下一次化装舞会上，款待我的朋友们，你可以演示一下你怎么做到的——我们会笑死的！"

……这仅仅是开始！……

"Look, you are a young frog and still have to learn a lot. I tell you, I draw water into my butt, take out the air I need to breathe, and then expel the water. That pushes me forward."

"Now, if you would like to entertain my friends at our next dress up party, you can show us how you do that – and we will have the greatest laugh ever!"

... AND IT HAS ONLY JUST BEGUN!...

……这仅仅是开始!……

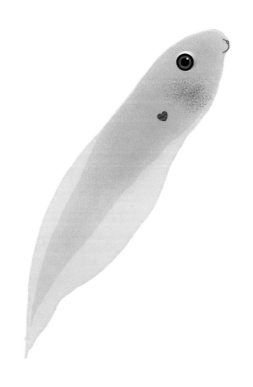

…AND IT HAS ONLY JUST BEGUN!…

# Did You Know?
## 你知道吗?

In Greek mythology a nymph is a natural spirit that looks like a beautiful maiden, living in rivers or woods. In nature it is an immature form of a dragonfly.

在希腊神话中，水虿是居于山林水泽的仙女，而实际上，它是蜻蜓的幼年阶段。

The dragonfly will spend 90%-95% of its life as a nymph, a development that can take from a few weeks to up to 3 years. Dragonfly larvae may moult 6 to 15 times.

蜻蜓生命里 90%—95% 的时间是以水虿的形态度过的，这个阶段短则几周，长的可达三年。水虿可能会经历 6 至 15 次蜕皮才能成为蜻蜓。

Dragonflies undergo an incomplete metamorphosis, skipping the pupal stage. It takes the nymph three hours to turn into a dragonfly by crawling out of the water, losing its mouth part, gaining wings by pumping liquid into their frames, and sitting in the sun to dry.

蜻蜓经历了一个不完全变态过程，跳过了蛹阶段。水虿需要花三个小时时间从水中爬出变成蜻蜓，在这个过程中，它丢掉了口器，向翅膀框架注入液体，待在阳光下晒干获得翅膀。

There are nearly 4 000 species of frogs. There are 5 000 types of dragonflies. It is difficult to distinguish them in their tadpole or nymph stage.

青蛙有将近4000种。蜻蜓有5000种。在青蛙和蜻蜓的蝌蚪或幼虫阶段，很难区分它们。

Some frogs lay their eggs in a foamy mass cocoon on a tree branch hanging over a pond or a stream. When it rains, the foam drips down dropping tiny tadpoles into the water below. It may seem like it is raining tadpoles.

一些青蛙在池塘或小溪上方树枝上的泡沫状的茧上产卵。下雨时，细小的蝌蚪从泡沫上往下滴到水中。看起来像是在下蝌蚪雨。

After hatching, the tadpole at first feeds off the yolk in its guts, and then it will feed on algae. Tadpoles will develop tiny tooth-like ridges that help grate food, plants in particular. Once they have legs, their diet includes insects.

刚孵出时，蝌蚪以其内脏里的卵黄为食，之后，它就以藻类为食。蝌蚪会长出细小的像牙齿一样的褶皱将食物摩擦成小颗粒，帮助进食。一旦蝌蚪长出腿来，它们的食谱中就包括昆虫了。

Frogs' tongues are attached to the front of their mouths rather than at the back, like that of humans. A frog catches an insect by casting its sticky tongue from its mouth and wraps it around the prey.

青蛙的舌头不像人类的舌头一样在嘴的后部,而是贴近嘴的前部,青蛙从嘴里吐出具有黏性的舌头,然后将它的猎物卷回来。

Frogs have a moist and slimy skin. Toads have a dry, bumpy skin that tastes bitter to animals trying to catch it. Their skin produces an odour that smells like that of a skunk. Toads can survive far from water. Toads run or make small hopping movements, while frogs take long, high leaps.

青蛙的皮肤潮湿而黏滑。蟾蜍的皮肤干燥且崎岖不平,让那些想捕食它们的动物感到苦涩。蟾蜍的皮肤产生一种类似臭鼬的气味,可以在远离水的地方生存。蟾蜍能够跑或者做小的跳跃动作,而青蛙则做长而高的跳跃。

# Think About It

## 想一想

Would you recognise every one of your friends at a dress up party?

你能在化装舞会上认出每一个朋友吗?

How would you respond if you recognise someone, but that person does not recognise you?

如果认出一个朋友,他却没认出你,你会有什么反应?

Who will be happy to have a chance, not just to dress up, but to transform your looks forever – with no way to return to your old look?

如果有机会永久改变外表,并且无法回到原来的模样(而不仅仅是化装),谁会为此感到高兴呢?

Do you like surprises? Or do you prefer that when things change you are quietly and carefully informed so as not to receive a shock?

你喜欢惊喜吗?或者,当事情变化时,你宁愿平静地、被小心地告知,以免感到震惊吗?

# Do It Yourself!
# 自己动手!

Dress up time! Find something to wear that your siblings and family have never seen. Make sure that it includes a wig and use make up that changes the colour of your face to one that no one associates with you. Also learn how to change your voice. How easy is it for you to confuse those who know you well, and make it impossible for them to recognise you? How successful are you at this? What do you think are the most effective ways to achieve this? Share your results with your friends.

化装时间到了！找一些你父母和兄弟姐妹从来没见过的东西穿上，一定要有一副假发，通过化妆，改变脸色确保没人会想到是你。也要学会怎样改变你的声音。看看能迷惑哪些非常熟悉你的人，让他们认不出来有多容易？你认为什么方法最有效呢？把你的结论与朋友们分享。

# TEACHER AND PARENT GUIDE

## 学科知识
### Academic Knowledge

| | |
|---|---|
| 生物学 | 青蛙是有脊椎冷血两栖动物；缺乏营养素可以导致失忆，特别是缺乏维生素$B_{12}$；蜻蜓的眼睛占据了它头部的大部分区域；成年的蜻蜓每日吃30至100只蚊子，是抑制蚊子种群的重要力量；为适应水中生存，水虿具有从水中分离空气和氧的能力。 |
| 化 学 | 胶原酶，一种降解胶原蛋白的酶，在脊椎动物、无脊椎动物和植物中广泛存在；在再生过程中一向以正效应而著称的抗氧化剂，在癌症治疗之后的组织再生过程中，却似乎具有负效应。 |
| 物 理 | 在类风湿性关节炎中软骨被破坏，减震性能消失，使病人的关节损坏并且疼痛；蜜蜂翅膀每秒钟震动300次，蜻蜓翅膀只震动30次；蜻蜓的眼睛有3万个复眼；青蛙的喉咙里有一个装满空气的小囊，囊中空气慢慢释放，发出了哇哇的叫声。 |
| 工程学 | 在蝌蚪尾巴里发现的软骨组织，含有很多蛋白聚糖，能起到减震作用；蜻蜓能直上直下地飞，并且能像直升机那样盘旋；蜻蜓用它们的脚捕捉猎物，命中率达到90%以上。 |
| 经济学 | 为了取得长远的成功，企业必须采取重大的变革，而且这种变革可以是破坏性的（约瑟夫·熊彼特），迫使公司里很多人适应新的环境，学习新的专业，调整自我，这就是如同蝌蚪变成青蛙或水虿变成蜻蜓一样的根本的变革。 |
| 伦理学 | 生活是面临不断变化的挑战；人们通常不愿意或者害怕改变，并试图强迫一切事物都保持现状，但是为了生存和发展，我们必须适应调整和转变。 |
| 历 史 | 蜻蜓大概是第一种有翼昆虫，它们大约存在了3亿年；历史上，蜻蜓象征成熟和深沉的性格。 |
| 地 理 | 蜻蜓可以来回跨越印度洋，在所有昆虫中迁徙距离最长（11000英里）；"变形"来自希腊语，意思是"改变"。 |
| 数 学 | 猎物判断蜻蜓并不是静止的唯一方法是看随着捕食者接近它体型的大小变化，这个问题是摄影几何学的一部分，摄影几何学是在几何透视研究历史中发展起来的一门数学学科。 |
| 生活方式 | 万圣节前夕是世界各地孩子们最喜欢的化装时间；使用一种流行的玩具"伪装机器人"（也被称为变形金刚），人可以变成机器。 |
| 社会学 | 温水煮青蛙的故事（查尔斯·汉迪）讲的是，尽管一个人面临生命威胁，必须要采取主动行动来改变现状，但却无动于衷，故步自封；面对不同的情况，我们需要适应不同的社会角色。 |
| 心理学 | 人从幼儿阶段起就喜欢穿各种服装：摆脱日常生活变成一个牛仔、公主或者老虎；因为穿着变化，自我感觉和行为习惯也发生改变。 |
| 系统论 | 在生活中变革是必需的，然而，人们倾向宁可改变周围的事物，而不是使自己适应新的环境，如年龄增长和职业改变。 |

# 教师与家长指南

## 情感智慧
### Emotional Intelligence

**青蛙**　青蛙是池塘里快乐的居民。他很执着，哪怕是在水蛭不仅不认识他，而且拒绝与他交谈的时候。他很自信，多次被拒绝却不受影响，坚持追问水蛭一点儿都不认识他的原因，以至于让水蛭感受到了威胁。青蛙寻找唤起水蛭美好回忆的方法，两次坚持说他真的就是那只蝌蚪。当被指控为食肉动物时，青蛙既没有承认也没有驳斥，而是通过问水蛭变形（通过变形，水蛭变成了有着长相不同，还有四个翅膀的小生物）的问题转移话题。当得知水蛭如何呼吸和在水中游动时，青蛙从吃惊转化为开玩笑。

**水蛭**　水蛭希望青蛙与她保持距离，因为水蛭将青蛙视为安全威胁。她坚持要独自待着，并且要求青蛙不要装着好像见过面、是老熟人，试图进行交谈。水蛭清楚青蛙既然变成食肉动物，他就具有食肉动物的危险，很可能将自己变成一顿美餐。她仔细分析了青蛙的变形，她了解了青蛙从蝌蚪转化成青蛙的变形过程之后，并没有显示出一点热情。当青蛙指出她将从水蛭变成蜻蜓之后，她依然无动于衷。水蛭不带感情、不谈细节，说明她的想法。一个细节（用屁股呼吸）并没被水蛭视为非凡，水蛭要求青蛙学习更多大自然的知识；当青蛙认为这个细节非常滑稽时，水蛭保持了她原有的方式，不动声色。

## 艺术
### The Arts

青蛙发出"哇哇"的叫声。不同的青蛙叫声区别很大，听起来可能像"唧唧"声、口哨声、"呱呱"声、"哞哞"声、"吱吱"声、"咯咯"声、狗叫声和"咕咚"声。现在听一听这些不同的声音，试试模仿一下。当几种不同的蛙叫声同时响起时，听起来像什么？这样听起来会很和谐，还是会很刺耳呢？

# TEACHER AND PARENT GUIDE

## 思维拓展
### Systems: Making the Connections

我们周围的事物总是在变化的——以至于我们的认识无法跟上那些正在发生的变化。挑战在于人们喜欢维持现状，做出极大努力并花费大额资金，让所有事情都保持原状——更有甚者，把事情弄得比原先看起来更好。甚至对身体进行整形手术使我们衰老的外表看起来年轻，或者造出理想的容貌。自然界展示了令人困惑的调整、改变、转化、适应甚至变形的能力。自然界很多物种知道在必要时如何隐藏、化装和伪装。通过这种方式，它们以不同的形式欢度它们的生活，这些改变和变形策略是它们维持持续适应能力的一部分，是确保它们获得更多水和食物有效的方式，并且维持物种已获取的在生态系统中的位置。人类需要认识到，我们是周围正在发生变化的原因，虽然我们应该对环境、生态和气候变迁负部分责任，但是我们却抵制改变我们自己。我们需要努力避免摧毁人类赖以生存的条件。第一步是理解我们周围每件事物是如何运行的，虽然一些科学发现令人惊奇，甚至看起来奇怪或好笑，人类应该深刻理解我们周围生命系统用于应对变化独具匠心的方法。它们利用温度、湿度、气压、食物可得性等外部变化，追求进化和共生的路径，最终实现生命的华章。如果人类要生存，那么我们就要改变我们的生活、生产和消费方式。

## 动手能力
### Capacity to Implement

列出自然界最惊人的十大变形。你知道了蝌蚪和水蚤的故事，但是还有更多。我们不应该只看物理变化，还要看生物物种用化学方法表演的惊人的特技，比如，大马哈鱼必须使其能够从在淡水中生活，转到海水中生活，然后再到淡水中生活。人也是不错的变形者：出生于纯液体的细胞，在体内生长出骨骼，当年龄增长时，骨头长得更长，更粗。

# 教师与家长指南

## 故事灵感来自

## 玛丽·施韦策
## Mary Schweitzer

玛丽·施韦策是一位古生物学家，她起初在犹他大学主修传播学，后来在蒙大拿大学获得第二教育证书，并获得生物学博士学位。她是第一个从恐龙化石中辨认和分离软组织（胶原）的研究者，她注意到在实验室降解的化学品，在受保护的自然环境中并不降解。施韦策还发现铁粒子在保存软组织的过程中可能起一定作用。科学家非常热衷于弄懂蝌蚪的尾巴是如何转变的，以及这一过程是如何在自然酶的作用下发生的。玛丽·施韦策还发现了恐龙和鸡之间的分子相似性，为二者之间的关联提供了进一步的证据。

图书在版编目（CIP）数据

化装舞会：汉英对照/（比）冈特·鲍利著；（哥伦）凯瑟琳娜·巴赫绘；何家振译. — 上海：学林出版社，2017.10
（冈特生态童书. 第四辑）
ISBN 978-7-5486-1265-0

Ⅰ.①化… Ⅱ.①冈… ②凯… ③何… Ⅲ.①生态环境－环境保护－儿童读物－汉、英 Ⅳ.① X171.1-49

中国版本图书馆 CIP 数据核字（2017）第 143501 号

ⓒ 2017 Gunter Pauli
著作权合同登记号　图字 09-2017-532 号

# 冈特生态童书
## 化装舞会

| | |
|---|---|
| 作　　者—— | 冈特·鲍利 |
| 译　　者—— | 何家振 |
| 策　　划—— | 匡志强　张　蓉 |
| 责任编辑—— | 许苏宜 |
| 装帧设计—— | 魏　来 |
| 出　　版—— | 上海世纪出版股份有限公司 学林出版社 |
| | 地　址：上海钦州南路 81 号　电话／传真：021-64515005 |
| | 网　址：www.xuelinpress.com |
| 发　　行—— | 上海世纪出版股份有限公司发行中心 |
| | （上海福建中路 193 号　网址：www.ewen.co） |
| 印　　刷—— | 上海丽佳制版印刷有限公司 |
| 开　　本—— | 710×1020　1/16 |
| 印　　张—— | 2 |
| 字　　数—— | 5 万 |
| 版　　次—— | 2017 年 10 月第 1 版 |
| | 2017 年 10 月第 1 次印刷 |
| 书　　号—— | ISBN 978-7-5486-1265-0/G.491 |
| 定　　价—— | 10.00 元 |

（如发生印刷、装订质量问题，读者可向工厂调换）

# 冈特生态童书（109-144）

水　　食物　　教育　　居住　　健康　　工作　　能源

记得要和身边的小朋友分享环保知识哦！
八喜冰淇淋祝你成为环保小使者！

联合国开发计划署支持

## 永远不要停止梦想

你拥有的梦想越多,你能重温的梦想就越多,实现梦想就越容易……永远不要停止梦想!

冈特·鲍利是世界零排放研究创新基金会(ZERI)的创始人,该基金会的理念是:"不要奢望让地球为我们创造更多资源,而要更善于利用地球现有的资源。"《冈特生态童书》正是这一理念的结晶。书中的一系列童话,看似充满奇思怪想,但却揭示了现实生活中的许多道理。孩子们通过这些童话学习科学知识,开发情商,提高艺术表达能力、动手能力和对复杂系统的理解能力。他们将成为真正的地球保护者。孩子们的内心充满着梦想和希望,千万不要让这些梦想和希望落空。未来一定会更加美好,因为孩子们的优秀总是超乎父母的想象。

上架建议:教育

定 价:10.00元
易文网:www.ewen.co

中国少年儿童生态意识教育丛书
环境保护部宣传教育中心　主编

冈特生态童书
118

ZERI "To never stop dreaming"
"永远不要停止梦想"

# 肥美的鸡肉

## Big, Fat, Tasty Chicken

**Gunter Pauli**
冈特·鲍利　著
凯瑟琳娜·巴赫　绘
颜莹莹　译

www.xuelinpress.com

# 亲爱的小朋友们

小朋友们，你们知不知道蘑菇也喜欢听莫扎特？知不知道山羊除了爱吃草，还会爬树、喜欢吃苹果、梨和栗子？冈特·鲍利先生为你们创作了许多有趣的童话故事，带着你们去奇妙的生态王国寻宝，相信在这段神奇的旅程中，你们一定会大开眼界。

如果读完故事还意犹未尽，"你知道吗？"会告诉你们更多的科学知识，"自己动手"会指导你们开辟自己的一片天地。最后，邀请你们的爸爸妈妈和老师们一起加入奇妙之旅吧！"教师和家长指南"为他们准备了各种背景知识，当然，遇到困难时，还是要请他们向工具书或者专家求助哦！

扫一扫，就能听冈特小故事。
更多惊喜，等你发现哦！